The Essential Guide On How To Raise And Keep Uromastyx Lizards: Care, Behavior, Diet, Feeding, Breeding, Housing, Handling, And Health Maintenance

Raymond S.L Gregory

Table of Contents

CHAPTER ONE 3
- UROMASTYX LIZARDS 3
- GEOGRAPHIC DISTRIBUTION OF UROMASTYX 6

CHAPTER TWO 10
- HABITAT AND ENVIRONMENT 10
- PHYSICAL CHARACTERISTICS AND MORPHOLOGY 13
- BEHAVIORAL TRAITS AND ADAPTATIONS 16

CHAPTER THREE 19
- DIET AND FEEDING HABITS 19
- REPRODUCTION AND LIFE CYCLE 21
- CHOOSING AND SETTING UP AN ENCLOSURE 24

CHAPTER FOUR 28
- TEMPERATURE, LIGHTING, AND STICKINESS NECESSITIES 28
- SUBSTRATE AND NATURAL IMPROVEMENT 31
- HANDLING AND SOCIALIZATION TIPS 34

CHAPTER FIVE 39
- HEALTH ISSUES AND CARE PRACTICES 39
- BREEDING UROMASTYX 42

CHAPTER SIX 45
- UROMASTYX BEHAVIOR 45
- HIBERNATION AND OCCASIONAL CHANGES 47

THE END 51

CHAPTER ONE
UROMASTYX LIZARDS

Uromastyx lizards, otherwise called spiny-tailed lizards, are captivating reptiles found in the rough deserts of North Africa, the Middle East, and portions of Focal Asia. They're prestigious for their spiked tails, which they can use as a guard instrument. Their name, Uromastyx, really signifies "tail (uromas) whip (tyx)" in Greek, demonstrating this particular feature.

These lizards are basically herbivores, feeding on different plants, flowers, and seeds. Their capacity to flourish in bone-dry conditions is amazing; they've adjusted to moderate water and can frequently go for extended periods without drinking. All things considered, they get dampness from their food and the climate, requiring an urgent transformation for endurance in their brutal habitats.

Uromastyx arrive in a range of varieties, from shades of brown and tan to lively yellows, oranges, and reds. Their hue can assist them with mixing into their environmental factors or, at times, act as an advance notice to expected predators.

As pets, Uromastyx are famous because of their tough nature and generally low upkeep requirements. Be that as it may, they really do require explicit care, incorporating a legitimate setup with a heat source, UVB lighting, and a reasonable diet to guarantee their prosperity. Giving them a balanced diet rich in greens, vegetables, and periodic fruits is fundamental to their health.

They're known for their lolling behavior, investing energy under a heat lamp to control their body temperature. This is pivotal for their processing and, generally speaking, metabolic capabilities.

Uromastyx are, for the most part, lone creatures and can be very regional, so it's prescribed to house them separately to forestall hostility.

Their one-of-a-kind appearance, fascinating behaviors, and moderately manageable care necessities make them captivating and remunerating reptiles for devotees and pet owners alike.

GEOGRAPHIC DISTRIBUTION OF UROMASTYX

Uromastyx are principally found in North Africa, the Middle East, and portions of south-central Asia. These interesting reptiles occupy dry and semi-parched locales, leaning toward deserts, rough landscapes, and sandy regions with restricted vegetation.

In North Africa, they are predominant across nations like Morocco, Algeria, Libya, Egypt, and Sudan. The Sahara Desert fills in as a huge habitat for different Uromastyx species because of its dry, blistering climate and immense spread of sand ridges.

Moving into the Middle East, Uromastyx can be found in nations like Saudi Arabia, Iraq, Iran, and portions of the Bedouin Promontory. They flourish in these locales because of the comparable desert-like circumstances and rough scenes that provide appropriate concealing spots and tunnels.

South-Focal Asia, especially in nations like Pakistan and Afghanistan, additionally has populations of Uromastyx, where they adjust to the dry conditions predominant here.

Their conveyance inside these locales can change in light of the particular species and their environmental necessities. These lizards are very much adjusted to make due in brutal conditions, using their tunneling skills to get away from outrageous temperatures and sidestep predators.

Their diets fundamentally comprise vegetation, seeds, flowers, and periodic bugs, and they have specific kidneys that permit them to limit water misfortune, a pivotal transformation for endurance in these dry habitats.

The geological dispersion of Uromastyx mirrors their exceptional capacity to flourish in probably the most outrageous conditions on the planet. Nonetheless, their populations face threats because of habitat obliteration, overgrazing, and assortment for pet exchange, highlighting the significance of preservation endeavors to safeguard these special reptiles in their local habitats.

CHAPTER TWO

HABITAT AND ENVIRONMENT

Uromastyx basically possess bone-dry and semi-parched districts of North Africa, the Middle East, and portions of Focal Asia. These terrestrial reptiles lean toward a habitat portrayed by rough deserts, sandy regions, and scrublands. They flourish in conditions with sweltering, dry climates, frequently in regions where temperatures can arrive at outrageous highs during the day and drop essentially around evening time.

Their environmental factors ordinarily comprise rough outcrops, tunnels, and fissures, offering shelter and insurance from predators and the cruel weather. Uromastyx are diurnal creatures, meaning they are dynamic during the day, basking in the sun to direct their body temperature and looking for shelter in their tunnels or obscure spots when temperatures take off.

The vegetation inside their habitat frequently incorporates tough desert plants like bushes, prickly hedges, and intermittent grasses. These plants give both food and shelter to the uromastyx. Their diet principally contains an assortment of vegetation, including leaves, flowers, and seeds. Water sources may be scarce in their current circumstances, so they've adjusted to get dampness from the plants they polish off, decreasing their reliance on ordinary drinking.

The spiny-tailed reptile's capacity to endure dry circumstances is because of particular transformations. Their thick, spiky tail supports thermoregulation, engrossing and delivering heat, while their scales and skin assist with decreasing water misfortune. Moreover, their tunneling behavior assists them with getting away from outrageous temperatures and threats from predators.

Human activities like habitat annihilation, overgrazing, and assortment for the pet exchange present threats to their populace nature. Preservation endeavors frequently center around safeguarding their natural habitats and controlling the exchange of these reptiles to guarantee their endurance.

PHYSICAL CHARACTERISTICS AND MORPHOLOGY

Uromastyx displays particular physical characteristics and a vigorous morphology. These reptiles are characterized by their strong, flattened bodies and short appendages. They normally range from 10 to 18 creeps long, with certain species arriving at up to 30 inches. Their bodies are shrouded in little, pointed, fell scales that give them a spiky appearance, particularly obvious in the tail, which is wide at the base and tightens to a sharp end, fixed with imposing spikes or thistles.

Their tinge changes among species, however, frequently incorporate lively shades like reds, oranges, yellows, and greens, which help in disguise inside their bone-dry habitats. Uromastyx have solid, strong heads with strong jaws, furnished with lines of little, sharp teeth adjusted for their herbivorous diet, which essentially consists of vegetation like leaves, flowers, and seeds.

These desert-staying lizards have an inclination for rough, sandy territories and can frequently be found in tunnels or clefts to get away from the outrageous heat of their current circumstances. Their bodies are intended for this brutal setting; their thick, layered skin forestalls water misfortune, and their nasal entries can proficiently hold dampness from breathed-out air, lessening water misfortune in the dry desert environment.

Their appendages are adjusted for digging and crossing lopsided ground, areas of strength for the feet, and strong leg muscles. The tail fills different needs, going about as a protection component and supporting balance and temperature regulation by retaining and scattering heat.

BEHAVIORAL TRAITS AND ADAPTATIONS

Uromastyx showcases different behavioral traits and transformations to flourish in its dry, desert-like habitats. These creatures are principally herbivorous, feeding on vegetation like leaves, flowers, and seeds. Their behaviors are appropriate to the difficulties of their current circumstances.

One of their unmistakable transformations is their capacity to endure high temperatures. Uromastyx loll in the sun to manage their body temperature, frequently seen relaxing on rocks to retain heat, then, at that point, retreating to tunnels or concealed regions to abstain from overheating. This behavior supports their thermoregulation, which is urgent for their endurance in the desert climate.

Their tunneling abilities are fundamental for getting away from outrageous temperatures and dodging predators. Uromastyx dig complicated tunnels, utilizing serious areas of strength for them and hooks, creating a place of refuge from the toxic components. These tunnels additionally act as retreats during cold evenings and breeding locales.

As far as guards, their tails assume an indispensable role. The tails of uromastyx are canvassed in spiked scales, filling in as a protection component against predators. When threatened, they can use their vigorous tails as a whip-like safeguard, preventing expected threats.

Socially, while they're for the most part single creatures, during the breeding season or in ideal habitats, they could connect, primarily for mating or laying out strength. Their communication includes body stances, head swaying, and variety changes to pass strength or status on to mate.

Uromastyx have an excellent water protection capacity. They extract dampness from their food and have productive kidneys that limit water misfortune, permitting them to get by in conditions with restricted water sources.

CHAPTER THREE

DIET AND FEEDING HABITS

Uromastyx have explicit dietary necessities that are pivotal for their health. These herbivorous reptiles dominatingly feed on a tight eating routine of new greens, vegetables, and a few fruits. Their meals basically comprise of mixed greens like collard greens, mustard greens, dandelion greens, and kale, giving them fundamental supplements like calcium and vitamins.

It's vital to try not to feed them high-protein foods like bugs, as these can cause health issues and are not part of their natural diet. Monetarily arranged diets made explicitly for herbivorous reptiles can likewise be remembered for their feeding schedule.

Uromastyx require a balanced diet to keep up with their health and forestall illnesses like metabolic bone infection, which can result from an absence of legitimate nourishment, specifically calcium and vitamin D3. Tidying their food with calcium and vitamin supplements is frequently prescribed to guarantee they get the important supplements.

Feeding habits ought to consider their age and size. More youthful uromastyx may require more regular feeding, while adults could have an alternate feeding plan. Giving new water is vital, although these reptiles may not hydrate as often as possible because of their desert-staying nature. Notwithstanding, a shallow dish of new water ought to constantly be accessible.

Understanding their natural habitat is fundamental for their diet and feeding. Uromastyx, being desert tenants, are utilized in a diet with a high fiber content and low dampness. Consequently, their diet ought to duplicate these circumstances as intently as could really be expected.

REPRODUCTION AND LIFE CYCLE

Uromastyx have an entrancing life cycle revolved around reproduction and endurance. These reptiles ordinarily arrive at sexual development between 2 and 5 years old, contingent upon the species and natural circumstances. The mating season for uromastyx generally happens in the spring, corresponding with expanded sunlight and hotter temperatures.

During this time, male uromastyx show regional behavior, vieing for the consideration of females. Males show strength by bouncing their heads, puffing their bodies, and now and again engaging in push-and-push communications. When a male effectively courts a female, sexual intercourse happens.

Female uromastyx lay grips of eggs, generally going from 7 to 35, in tunnels or shallow homes dove in sandy or loamy soil. The eggs are then covered and left to hatch, profiting from the glow of the climate. The hatching time frame goes on for around 70 to 90 days, contingent upon factors like temperature and mugginess.

Upon hatching, the youthful uromastyx arises. They are free from birth and should fight for themselves. They quickly look for shelter, warmth, and food. These youthful lizards principally feed on vegetation, which consists of leaves, flowers, and seeds.

Their development rate is somewhat sluggish, and it requires a couple of years for them to arrive at development. As they develop, uromastyx shed their skin intermittently, taking into consideration legitimate development and transformation to their evolving climate.

All through their lives, uromastyx face different difficulties, including predation, ecological changes, and a contest for assets. Nonetheless, these lizards have adjusted instruments to make due, including their capacity to manage body temperature, look for shelter, and depend on a fundamentally herbivorous diet.

CHOOSING AND SETTING UP AN ENCLOSURE

Choosing and setting up an enclosure for a uromastyx requires careful thought to impersonate its natural habitat. Right off the bat, pick a terrarium that is sufficiently roomy, something like 40 gallons for a solitary uromastyx, with great ventilation and secure terminations. A bigger enclosure is useful, giving adequate space for movement.

Substrate choice is crucial. Use materials like reptile rugs, sand, or a sand-soil blend. Try not to utilize free substrates that could cause impaction whenever ingested. Incorporate concealing spots like rocks, logs, or artificial caverns to create a feeling that all is well with the world for your uromastyx.

Heat and light are essential for their prosperity. A luxuriating spot temperature ought to be between 100 and 110°F (37 and 43 °C), and the cooler side around 80°F (27°C). An inclination permits them to control their body temperature. Utilize a heat source like a heat light or ceramic heat producer, and utilize an UVB bulb to support their vitamin D3 union, which is fundamental for calcium retention.

Keep up with moisture levels around 20–40%. Uromastyx are desert occupants, and high moisture can prompt health issues. A shallow water dish is adequate for drinking and incidental dousing.

Diet is a huge viewpoint. Uromastyx basically eats vegetation. Offer various new greens, vegetables, and intermittent fruits. Collard greens, dandelion greens, and ringer peppers are great decisions. Calcium and vitamin supplements ought to be given to guarantee they get legitimate sustenance.

Consistently perfect the enclosure to keep up with cleanliness, eliminating dung and uneaten food. Play out a profound clean month to month to forestall the buildup of microorganisms.

It's vital to screen your uromastyx for any indications of stress, an ailment, or unusual behavior. Guarantee customary check-ups with a reptile veterinarian to keep up with their health.

CHAPTER FOUR

TEMPERATURE, LIGHTING, AND STICKINESS NECESSITIES

Uromastyx are desert-staying reptiles with explicit natural requirements. Keeping an ideal climate for these creatures is pivotal to their prosperity.

Temperature: Uromastyx require a warm climate as they begin in sweltering desert locales. The relaxing spot in their enclosure ought to stretch around 100–110°F (37–43°C) during the day, while the cooler end ought to associate with 80–90°F (27–32°C). Around evening time, the temperature can decrease to around 70–75°F (21–24°C). Utilizing heat lights or ceramic heaters can help achieve and keep up with these temperatures.

Lighting: Uromastyx need openness to UVB light to impersonate the natural daylight they'd get in nature. UVB lighting assists them with processing calcium and prevents issues like metabolic bone infections. A full-range UVB bulb ought to be accommodated for around 10–12 hours every day to guarantee they get the fundamental UV openness.

Mugginess: These lizards come from bone-dry conditions and flourish in low dampness. Their enclosure ought to be somewhat dry, with a dampness level in a perfect world around 20–30%. Extreme dampness can prompt respiratory issues and skin issues for uromastyx. Giving a shallow water dish to them to drink from is adequate; notwithstanding, keeping up with dry substrate and great ventilation is vital to forestall overabundance dampness.

Keeping up with these circumstances requires exact thermometers and hygrometers inside the enclosure. The actual enclosure ought to be adequately open to take into consideration temperature slopes, with hides or shelters provided at both the warm and cooler ends, empowering the uromastyx to manage their body temperature depending on the situation.

Ordinary checking and changes are imperative to guarantee the prosperity of uromastyx. Giving a balanced diet, proper shelter, and a reasonable climate regarding temperature, lighting, and mugginess are key components in caring for these unbelievable desert-staying reptiles.

SUBSTRATE AND NATURAL IMPROVEMENT

Uromastyx flourish in conditions that recreate their natural habitat. The substrate, the material covering the enclosure floor, is significant for their prosperity. A blend of sand and soil functions admirably, imitating their desert starting points. It ought to take into consideration tunneling and natural behaviors. Stay away from substrates that could be ingested and cause health issues, as well as free particles that might affect their processing.

Ecological advancement is key to their psychological and physical health. Giving them different concealing spots, shakes, and branches permits them to investigate, climb, and relax under heat sources. Artificial plants and designs that look like their natural climate can offer mental excitement and copy their wild habitat, advancing their prosperity.

It is imperative to keep an ideal temperature. Uromastyx incline toward a hot climate, with a luxuriating spot stretching around 100–110°F (37–43°C) and a cooler side around 80°F (27°C). Heat sources like heat lights or heat cushions should be carefully situated to create these temperature slopes while guaranteeing no immediate contact that could cause consumption.

UVB lighting is vital for their calcium metabolism and overall health. Natural daylight or specific UVB bulbs are important to support calcium ingestion and forestall metabolic bone infection.

It is similarly critical to maintain a balanced diet. Uromastyx fundamentally eat vegetation, so a diet rich in salad greens, vegetables, and periodic fruits is great. Keep away from high-protein diets that could prompt health issues.

Consistently perfect and keep up with the enclosure to forestall bacterial development and keep a healthy climate for your Uromastyx. Spot cleaning, eliminating uneaten food, and consistently changing the substrate are fundamental.

Noticing their behavior and health is vital. Any indications of stress, like a loss of craving or surprising behavior, ought to provoke a vet visit.

HANDLING AND SOCIALIZATION TIPS

1. Start Sluggish: Uromastyx could not at first appreciate handling. Start by essentially being close to their enclosure, permitting them to get familiar with your presence.

2. Gradual Introduction: When they appear to be agreeable, begin with brief and delicate handling meetings. Start by

petting them delicately while they're in their enclosure.

3. Secure Climate: Consistently handle them over a delicate surface and near the ground to keep wounds from falling. Uromastyx could wriggle or jump when handled, so a protected grasp is fundamental.

4. Consistent interaction: Normal, short-handling meetings assist them with acclimating to being held. Be patient and steady in your methodology.

5. Respect Limits: Uromastyx could show pressure signals like puffing up or tail-whipping. Assuming they appear to be troubled, give them space and attempt something else.

6. Observe Body Language: Watch for indications of distress or stress. On the off chance that they appear to be fomented, set them back in their enclosure and have a go at handling them later.

7. Positive Support: Offer treats subsequent to handling to create positive affiliations. This aides in building trust and diminishing their pressure.

8. Routine and Persistence: Lay out an everyday practice for handling to assist them with having a solid sense of safety and anticipate their cooperation. Persistence is vital; each reptile changes at its own speed.

9. Limit Unpleasant Experiences: Stay away from clear commotions or abrupt movements around them, as these can pressure the uromastyx. A quiet climate helps with their socialization.

10. Professional Direction: In the event that you're unsure or the uromastyx appears to be determinedly worried, look for an exhortation from a reptile-smart veterinarian or an accomplished reptile handler.

Keep in mind that uromastyx have unmistakable characters, so some could take more time to acclimate than others. Continuously focus on their solace and prosperity. With persistence and a steady, delicate connection, numerous uromastyx can turn out to be more lenient toward handling over the long haul.

CHAPTER FIVE

HEALTH ISSUES AND CARE PRACTICES

Uromastyx are desert reptiles requiring explicit care to remain healthy. Key health issues for these lizards frequently originate from deficient farming and natural circumstances. One common issue is metabolic bone sickness (MBD), which comes about because of an absence of proper UVB openness and a lack of calcium. This prompts debilitated bones and can be lethal on the off chance that it is not tended to. Giving an UVB light source and a calcium-rich diet is critical to preventing MBD.

Another issue is the warm guidelines. Uros need a slope in their enclosure, with a hot lolling spot around 110–120°F and a cooler region around 80–90°F. Wrong temperatures can cause pressure and influence their processing and immune systems. Legitimate heating components, for example, clay heat producers or heat lights, are fundamental for keeping up with these temperatures.

Drying out is a huge concern. Uromastyx begin in dry districts and get a large portion of their water from their food. Hydration can be guaranteed by offering new greens, vegetables, and incidental fruits. A shallow water dish for drenching or drinking is likewise important; however, they may not utilize it as often as possible.

Parasitic contamination, especially interior parasites, can influence uromastyx. Customary veterinary check-ups and waste tests are indispensable for early discovery and treatment.

To guarantee their prosperity, keep a reasonable habitat with a substrate like sand and soil blended. Uros are burrowers, so giving hides and designs to climbing and luxuriating is fundamental for their psychological and physical health. Standard cleaning of the enclosure is important to forestall bacterial and parasitic development.

Noticing your uromastyx consistently for any indications of sickness, changes in behavior, or cravings is urgent. On the off chance that you notice anything unusual, counseling a reptile veterinarian is enthusiastically suggested. Careful regard for their current circumstances, diet, and behavior plays an urgent role in keeping up with the health and imperativeness of these fantastic reptiles.

BREEDING UROMASTYX

1. Habitat Setup: Create an open, exceptional habitat with legitimate heating, lighting, concealing spots, and a sandy substrate impersonating their natural climate.

2. Presenting the Pair: Acquaint the male with the female's habitat or give a nonpartisan space where they can collaborate. Screen their behavior to guarantee they don't become forceful.

3. Temperature and Lighting: Keep up with the ideal temperature and UV lighting in the habitat, repeating their natural circumstances. Satisfactory heating and lighting are vital for their regenerative health.

4. Courtship and Mating: Uromastyx show courtship behaviors, including head weaving, tail flicking, and orbiting. At the point when the male is acknowledged, mating happens, commonly during the hotter months.

5. Egg-Laying: The female will dig an opening in the substrate to lay eggs. Furnish a different compartment with a reasonable substrate for egg-laying in the event that they don't do it in the primary enclosure.

6. Hatching: Carefully gather the eggs and spot them in a hatchery with fitting temperature and dampness settings. Watch out for the eggs during the hatching time frame, which ordinarily lasts around 70–90 days.

7. Hatching and Care: When the eggs hatch, guarantee the infant Uromastyx has a reasonable setup looking like the grown-up enclosure. Give appropriate sustenance and care to the hatchlings.

CHAPTER SIX

UROMASTYX BEHAVIOR

Luxuriating: Uromastyx depend intensely on relaxing to control their body temperature. They spend a huge part of their day under a heat source, engrossing warmth to help processing and, by and large, health.

Tunneling: These lizards are capable diggers and look for shelter in tunnels to get away from outrageous temperatures and predators. In captivity, giving a substrate that permits tunneling mirrors their natural behavior and gives a feeling of safety.

Social Association: Uromastyx aren't regularly friendly creatures and may show regional behavior, particularly during mating seasons. In captivity, it's normal for different Uromastyx to live, respectively, yet careful consideration regarding their collaborations is important to forestall hostility.

Feeding Patterns: Their diet consists predominantly of vegetation, and they have an exceptional behavior of sometimes relaxing subsequent to eating. They are known to be shrewd feeders and can, at times, consume their shed skin, which could appear odd yet is a natural behavior.

Brumation: Uromastyx go through a time of diminished action, like hibernation, called brumation. This normally happens during the colder months and is fundamental for their general prosperity.

HIBERNATION AND OCCASIONAL CHANGES

Uromastyx are adjusted to survive cruel ecological changes through a cycle known as brumation, as opposed to genuine hibernation. During colder months or in outrageous heat, these reptiles experience a time of diminished action, metabolic log jam, and decreased food consumption. This brumation period permits them to ration energy and endure when assets are restricted.

As temperatures decrease, the uromastyx instinctually senses the change and begins planning for brumation. They look for shelter in tunnels or hideaways that offer assurance from the cruel weather. Inside these asylums, they become less dynamic, once in a while remaining torpid for quite a long time or even months. This decreased action assists them with monitoring energy by dialing back their metabolism.

During this time, their body temperature and metabolic rate decline altogether, permitting them to get by without devouring as much food. They may at times wake to hydrate, yet by and large stay in a state of rest.

When temperatures become better, uromastyx, bit by bit, rise out of their torpid state. They start to expand their action, looking for food and lolling in the sun to raise their body temperature and metabolic rate. This enlivening stage is critical for them to recover their solidarity and resume their ordinary exercises.

Uromastyx are versatile creatures, adjusted to make due in dry conditions where temperature changes are normal. Their capacity to brumate assists them with getting through cruel circumstances, guaranteeing their endurance when assets are scarce. Be that as it may, the particulars of this interaction can change among individual uromastyx in light of elements like species, age, and generally health.

To support these reptiles during their breeding periods, it's fundamental to provide a reasonable habitat that impersonates their natural climate, with sufficient shelter and temperature inclinations. Moreover, keeping a legitimate diet and guaranteeing access to water during brumation are essential for their prosperity and effective progress all through this lethargic stage.

THE END

Made in the USA
Columbia, SC
13 May 2024

35606934R00030